REPORT

ON THE

GEOLOGICAL SURVEY

OF THE

STATE OF VERMONT,

BY

EDWARD HITCHCOCK,

STATE GEOLOGIST.

IN SENATE, Nov. 2, 1858.

Communicated by the Executive, laid on the table; and, *Ordered*: That 1000 Copies be printed for the use of the General Assembly.

(Attest,) CLARK H. CHAPMAN, Secretary.

BURLINGTON:
DAILY TIMES OFFICE PRINT.
1858.

Hon. Burnham Martin,

President of the Senate.

Sir :

I have the honor to transmit to the Senate, for the use of the General Assembly, the Report of Rev. Edward Hitchcock, State Geologist.

HILAND HALL.

Executive Chamber,
November 2, 1858.

REPORT ON THE GEOLOGICAL SURVEY.

TO HIS EXCELLENCY, RYLAND FLETCHER,

Governor of Vermont.

SIR :—

Agreeably to my instructions, I have the honor to present my Second Annual Report, as State Geologist.

Like the last, this Report will be brief, and consist of a statement of what myself and assistants have been doing during the year, with suggestions as to the future. I have supposed it to be understood that we should not attempt to give any detailed description of the rocks of the State, till our Final Report. Our time has been occupied in obtaining materials for that Report, and we are now ready to enter upon its preparation.

Last year, I suggested the desirableness of pushing the Survey with so much energy, that the essential part of the explorations abroad should be completed by the present time; and I pledged myself that this should be done, if the annual appropriation were increased to fifteen hundred dollars. As this was not done, the question came up, Shall we, nevertheless, attempt to accomplish an object so desirable? We knew that the usual appropriation of one thousand dollars would probably have been entirely exhausted by our expenses, leaving nothing for services, either to myself or assistants. We determined, however, to take this course, not doubting that we might depend on the liberality and generosity of the State as to a compensation, provided its Legislators shall see evidence, in our reports and collections, that we have been faithful in the work. Two and sometimes three parties have, therefore, been in the field during most of the summer. We entered into an arrangement with the Rev. S. R. Hall, of Brownington, who was early associated with Professor Adams in the Geological Survey, to go over the three northern sections, giving special attention to the north-east part of the State, which had never been but imperfectly examined. For this service he is to receive the moderate sum of seventy-five dollars ; and as he is quite familiar with the geology of the State, especially its northern part, he was requested to extend his report as far south as he should choose, and present some views upon Agricultural Geology ; although that subject was not included in my commission.

The rest of us have been employed in reviewing and finishing the sections commenced last year, and in measuring several new ones ; so that fourteen are now completed. The specimens collected for the State Cabinet now amount to several thousands : the exact number we cannot specify, because the room in the new State House, where they are to be displayed, is yet unfinished ; and therefore, the specimens, though chiefly sent to Montpelier, are still in boxes. We have, also, been endeavouring to fix more accurately the outlines of the different rocks, on the Geological Map, and to look up new facts of economical importance.

We have, likewise, made some progress in that most difficult of all the problems in Vermont Geology, to determine what the rocks were before they were metamorphosed. It is not very difficult to give an appropriate name to these rocks in their present condition, but every careful observer finds that they have, almost without exception, undergone important changes; and the inquiry is, What were they before the change ? Or, to make the question, more specific, Are the non-fossiliferous rocks of Vermont merely the fossiliferous strata of New York metamorphosed ? And if so, Can we identify the varieties of the crystalline rocks of Vermont with those of the New York series ? It may seem of little practical importance to determine such points. Yet in fact, questions as to the useful substances that may be expected to occur in particular rocks, can sometimes be answered only by tracing them back through the mazes of metamorphism to their origin. We hope to be able to do this to some extent; but in some cases, with the light we now have, the identification will be little better than conjectural.

It will be needless to go into detail as to all the points to which our attention has been directed, during the year. The result of the whole is, that by our extra efforts we feel prepared to enter upon the preparation of a Final Report. We cannot, indeed, settle all the scientific questions that relate to the rocks ; but we hope to be able to give a fair view of them. So far as their economical value is concerned, we trust we have the materials for its full elucidation; and this we suppose to be the chief object of the survey. We hope, also, to present not a little that will be interesting to the enquiring and scientific.

I do not mean that there are no points in the Geology of the State which do not need further examination. We know of several such points, and of importance too, about which we are still very much in the dark. But the examination of those can be made while we are proceeding with the preparation of the Report. Other points, also, will doubtless suggest themselves as we get the specimens in their proper order, and as we transfer to paper the labours of the past summer.

The subjects that have received less attention than they deserve, are the following,—

1. *The Mineral Springs.*—I should judge there might be ten or twelve of those, important enough to demand an analysis. But analyses cannot be made without expense, and hitherto we have had no money to devote to the mineral springs. A part of their analysis (the determination of the gaseous ingredients,) must be made at the springs, and a part in the laboratory. I think it could be done for about $8 or $10 for each spring, exclusive of traveling expenses. If the Government wish this work done, the Chemist of the survey will see that it is accomplished, without delaying the preparation of the Report.

2. *Analyses of other substances.*—In the course of the survey not a few analyses have been obtained; but during my connection with it, no charge for them has been made, because I knew that no money could be spared for this purpose. It has been left, therefore, with the Chemist of the survey, to secure as many analyses as he felt willing to execute without compensation. But many substances remain without analysis, whose composition might be of great use, both economically and scientifically. I think that one or two hundred dollars would enable the Chemist to obtain the most important of these analyses.

3. *The scenery.*—I cannot believe that the citizens of Vermont will be willing to have a Geological Report appear, that does not contain at least a few sketches, of the fine scenery for which the State is so well known; especially when those sketches can be taken where they will afford fine geological as well as scenographical illustrations. We have not, however, employed any one to make such sketches, for the reason specified under the last head; although a competent artist is ready to do this work, charging only for his traveling

expenses. If fifty or seventy-five dollars were at our command, I think we might secure a least a dozen views, that would not only attract the tourist but the geologist.

I think, then, that 300 or 400 dollars would secure the accom plishment of the three objects above specified; I mean that sum as an extra appropriation. For I think the Government will see that the usual sum devoted to the survey will not be sufficient for the purpose. Whether the objects be worth the outlay, the Legislature must judge.

4. *Surface Geology.*—As the bearings of this subject are almost wholly scientific, we have not thought it proper to devote but very little time exclusively to it; but have, nevertheless, collected many facts in surface geology while at work upon things more essential. Yet the deficiences are numerous. We shall not be able, for instance, to map the terraces along but a part of the rivers; nor to give but a part of what we regard as sea beaches and sea bottoms, when the continent was under the ocean; nor to trace out only a few of the old river beds of a former continent. But we hope to present enough of all these and similar phenomena, to answer as specimens, and stimulate others to hunt up more. And we do not think it a sufficient reason for delaying our Report, because we might find a greater number by delay.

5. *The identification of metamorphosed rocks with known fossiliferous strata.*—I have already described the great difficulty of this work, yet we find, from year to year, that the leaves of the rocky volume are slowly opening, and the papyrus unrolling. I doubt not that, were the survey to be continued longer, each year would add something to our present imperfect knowledge. Yet perhaps the interest which we hope to awaken by what we shall present, may do as much as our continued labours.

6. *An examination of the regions bordering on Vermont.*—On the South, the Vermont rocks extend into Massachusetts and New York, and on the North into Canada. On the West and East the rocks of New York and New Hampshire are intimately connected with those of Vermont. It is fortunate that those of Canada, New York and Massachusetts, by means of the Geological Surveys that have been executed in those States, have been mapped. But, as every geologist knows, we onght to have opportunity to examine them for ourselves, in order to compare them with those of Vermont. Without such examination to some extent, indeed, no geologist would feel safe in deciding upon the character of the latter. The rocks of Canada, especially, ought to be examined; since the Ver-

mont rocks extend into that province and are there less metamorphosed. But we have not done it to much extent. Indeed, the only explorations we have made out of the State, have been, to pass once to the top of the White Mountains, carrying thither one of our sections; and extending another a few miles West of Lake Champlain. Had we the means, most gladly would we make these border explorations; and we doubt not that the Vermont survey would reap the benefit.

In spite of these and similar deficiencies, I judge that the most important objects of a Geological Survey, which the State has had in view, are now so far accomplished that we may set about arranging our materials for a Final Report, provided that, while engaged in this work, we may be allowed to go abroad, occasionally, to get more light on certain unsettled points, and resolve other questions that will undoubtedly arise, when we can see all the specimens arranged in their true places in the State Cabinet, and attempt to transform our field notes to Sections and Maps. Several years more of exploration would, indeed, be necessary, to enable us to make a Report entirely satisfactory to ourselves. But we trust that we are sufficiently masters of the Geology of the State not to do discredit to ourselves, or dishonor to the Commonwealth.

I might allude to more personal matters, as a reason for entering at once upon the preparation of our Report. It is certainly desirable to do it while the Geological Corps is unbroken. For it has been understood that each man should report upon some particular department of the survey, and this has turned the attention of each one to his specified department, and he can probably do it more justice than any other person. I know enough of the plans of some of our number, to suppose that they do not feel able to devote more than one year more to this work; and the frailty of my own physical system admonishes me not to make much calculation upon the future for the performance of labour.

It is presumed the Government will wish me to present as definite a description and outline of the Report which we propose to make, as I am able to do. The following synopsis will give an idea of the work, as it at present lies in my mind. I would reserve the right, however, to modify its subordinate parts, should further reflection make it desirable.

Scientific Geology.

1. A brief outline of the leading principles ef Geology; ex. gr., *a* Stratified and Unstratified Rocks; *b* Palæontology; *c* Surface Geology; *d* Metamorphosing rocks fully considered.

Rocks of Vermont.

1. *Stratified.*

1. Alluvium, or Surface Geology.

1. Drift.

a Common Drift; *b* Drift Striæ; *c* Embossed rocks; *d* Bowlders; *e* Crushed ledges; *f* Trains of angular Blocks.

2. Modified Drift.

a Ancient Beaches; *b* Terraces; *c* Moraine Terraces; *d* Old Sea Bottoms; *e* Ancient Glaciers, (their striæ and moraines); *f* Pond Ramparts; *g* Old river Beds; *h* Vallies of Erosion; *i* Clays; *j* Marl; *k* Peat; *l* Bog Iron Ore; *m* Wad, (manganese); *n* Fossil Marine Shells; *o* Fossil Cetacea; *p* Fossil Mammalia.

2. Tertiary Deposits: *a* Kaolin and coloured Clays; *b* Ochres; *c* Limonite, (brown hematite); *d* Manganese; *e* Fossil Fruits; *f* Brown Coal.

3. Silurian Rocks.

a Potsdam Sandstone; *b* Calciferous Sandrock; *c* Chazy Limestone; *d* Trenton Limestone; *e* Utica Slate; *f* Hudson River Group.

4. Metamorphic Rocks, schistose, slaty and massive, (probably Silurian and Devonian Rocks originally.)

a Gneiss; *b* Mica Schist; *c* Talcose Schist; *d* Argillaceous Slate; *e* Quartz Rock; *f* Hornblende Schist; *g* Soapstone; *h* Serpentine.

5. Hypozoic Rocks, (Laurentian); *a* Gneiss.

2. *Unstratified Rocks.*

1. Granite.

a Common; *b* Concretionary; *c* Conglomerated; *d* Porphyritic; *e* Graphic; *f* In Veins.

2. Syenite.

a Common; *b* Conglomerated and Brecciated; *c* In Dykes.

3. Porphyry.

a Granitic; *b* Trachytic.

4. Trap.

a Augitic; *b* Trachytic.

5. Dykes and Veins.

a Granitic; *b* Porphyritic; *c* Trappous; *d* Quartzose; *e* Calcareous; *f* Epidotic; *g* Metallic.

2. Economical Geology of the State, *or substances useful to the arts.*

a Limestone as a fertilizer; *b* as Cement; *c* as Marble; *d* Soapstone; *e* Serpentine; *f* Granite; *g* Flagging Stones; *h* Roofing Slate; *i* Iron; *j* Copper; *k* Lead; *l* Gold; *m* Chrome; *n* Nickel; *o* Titanium; *p* Manganese; *q* Iron Pyrites; *r* Ochres; *s* Clays; *t* Quartz; *u*

Feldspar; *v* Brown Coal; *w* Whetstones; *x* Hones; *y* Marl; *z* Peat; *aa* Mineral Springs.

3. Detailed Reports.

1. On the Sections generally. By Charles H. Hitchcock.

2. On the three Northern Sections, with applications to Agriculture· By Rev. S. R. Hall.

3. On the simple Minerals of the State. By Edward Hitchcock, Jr.

4. On the Chemistry of the Survey. By Charles H. Hitchcock.

5. On the Economical Geology. By Albert D. Hager.

As to the size of a Report drawn up according to this outline, I think it would form a quarto volume of some 5 or 600 pages, perhaps some less—possibly larger. I must be more indefinite as to the number of Plates. There must be, first, a Map of the General Geology; next, a Map of the Surface Geology; next, 14 or 15 Sections, occupying, say three or four plates; next, (if possible to obtain them,) 10 or 12 views of Scenery, two of which might generally be put upon a plate, making six plates more. If it should be thought best to figure all the organic remains found in the State, as well as many of the curious concretions in the newer deposits, it would require quite an addition to the plates. But I judge that if we increase the number already mentioned, (say twelve), to twenty, it would be sufficient. That number might be conveniently bound with the text into one rather thick quarto volume. Besides these plates, I think that probably 200 wood cuts would be required.

I have made these estimates in the belief that the Government would wish to know what sort of a Report we have been proposing to make out. A smaller one, with fewer illustrations, might, indeed, be prepared. But I have presumed that Vermont, taking a prominent place as it does among the States, in its subterranean resources, and in the scientific interest of its rocks, will not wish to fall behind them in the manner of exhibiting its geology. I would, indeed, practice a strict economy in this matter. But I cannot believe that it will be creditable to the State, or in accordance with the wishes of its enlightened citizens, to bring out a meagre and stinted account of its geology; and such I think it must be, if on a much less scale than the above estimate. I am unable to state the cost of getting out such a volume; but from the details given, I presume that those familiar with such matters could determine the expense approximately.

I have another object in this attempt at definite statement; viz. to ascertain the wishes of the Government. For if they do not approve of our plan, we shall be glad to know it, and thus be saved much useless labour.

As to the time requisite for the preparation of such a Report as I have indicated, much will depend upon health, and the greater or less number of unsettled points we may find requiring re-examination. But it would be unreasonable, if I may judge from experience, to expect that such a volume could be got ready for publication, (should the Government wish to publish it,) in less than a year. If, however, they choose to refer the question of the publication of the Report to a Committee, with power to proceed with the work as soon as a part of it is ready, if they should judge it worthy, the work might be got out several months earlier than otherwise. This course has been adopted once or twice by the Government of Massachusetts, in similiar cases, and with such a result as I have indicated.

Allow me to make a few suggestions as to the arrangement of the specimens, collected by us, in the State Cabinet. I have been shown the room in the new State House, which is understood to be appropriated to this object, and I have supposed that without doubt the specimens of rocks illustrating the Sections, would be placed upon the walls. If the South wall and the West wall, as far as the door, were to be used for this purpose, I would suggest the following arrangement: Along the line of the Sections, it may be remembered, the greater part of the specimens were collected. Suppose we begin with Section No. 1, near the bottom of the wall. Let a space, say six inches wide, be marked across the wall, leaving room enough beneath the strip for one, two or three shelves, on which, in the order in which they were collected, all the specimens obtained on that section may be placed. In the strip above them, let the section itself be painted, each rock having its proper thickness and dip. Section 2nd may be placed in like manner immediately above section 1, and one, two, or three shelves be put beneath for specimens, according to their number; and so on till all the sections are placed upon the wall, one above the other. The sketch below will give an idea of this arrangement. I have never

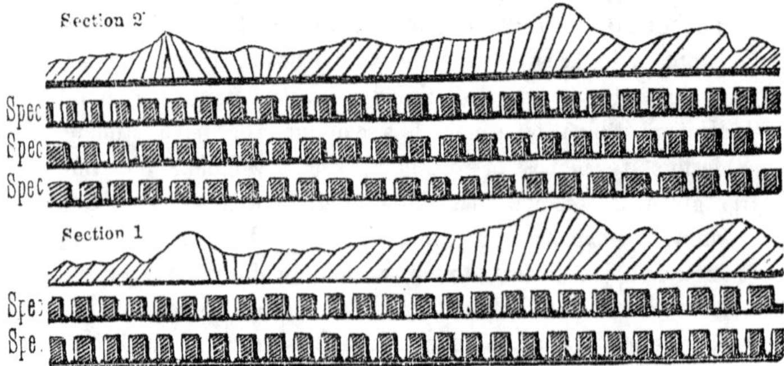

seen one like it; but it seems to me as if a cabinet, thus arranged, would give visitors, at a glance, an idea of the Geology of the State, and enable them at once to determine what the rock is, in any particular district.

I know not how soon the room in the State House, appropriated to the Cabinet, may be so nearly finished that we can display the specimens in it, even upon the floor. If not till next summer, it may delay our Report, since we need to examine the specimens, lying in their proper order, before we are able to describe them. or discover deficiencies. If they must be kept in boxes till next year, I beg leave to suggest the expediency of obtaining an insurance upon them, which would be some consolation, should they share the fate of the previous collection. I think it a moderate estimate to say that the cash value of the whole cannot be less than $1500; though it would cost more than that, probably, to obtain another like it.

Our expenses, the present year, up to this time, have amounted to *Seven Hundred and Thirty Seven Dollars and Twenty-five cents.* This leaves a balance of $262.75 for compensation. I divided it between my two assistants, (A. D. Hager and Charles H. Hitchcock,) whose whole time is given to the work, making *One Hundred and Thirty-one Dollars and Thirty-seven cents to each.*

I have spent over six weeks in the field, this season, but shall present no bill except for expenses. As to compensation for services, I shall leave it to the liberality of the State, when the work is done. At this time, I ask only that the annual appropriation of One Thousand Dollars, which the Act for completing the Survey directs to be paid, " until otherwise ordered by the Legislature," be continued another year.

With high respect,

Obediently yours,

EDWARD HITCHCOCK.

Amherst, Oct. 12th, 1858.

www.ingramcontent.com/pod-product-compliance
Lightning Source LLC
LaVergne TN
LVHW020643110826
845149LV00004B/1327

* 9 7 8 1 4 1 8 1 8 9 7 7 8 *